FUN WITH FOSSILS

AMMONITES

Left to right
1 *Dactylioceras commune*
2 *Microderoceras birchi*
3 *Arnioceras semicostatum*
4 *Amaltheus* species
5 *Androgynoceras* species

1 *Arietites rotiformis*
2 *Schlotheimia angulata*

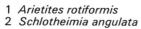

1 *Liparoceras cheltiense*
2 *Androgynoceras* species

1 *Kosmoceras jason*
2 *Nautiloid section*

William C. Cartner

FUN WITH
FOSSILS

KAYE & WARD · LONDON
in association with Methuen of Australia Pty. Ltd.
and Hicks, Smith & Sons, New Zealand

Mantelliceras mantelli 80 mm
Cretaceous ammonite

First published by
KAYE & WARD LIMITED
21 New Street, London EC2M 4NT
1970
Second impression 1977

ISBN 0 7182 1312 2

Printed in Great Britain by Whitstable Litho Ltd, Whitstable, Kent

CONTENTS

Teallocaris loudonensis 35 mm
Carboniferous shrimp
(Royal Scottish Museum Crown Copyright)

Promicroceras species—actual size
Jurassic ammonites

ACKNOWLEDGEMENTS

For some of the material used in this book the author gratefully makes acknowledgements to:

The Director and Staff, Teesside Museums Service

Shell International Petroleum Company Limited

The Northern Echo

The British Young Naturalists Association –Director G. G. Watson Esq.

J. Young Esq., The Hancock Museum, Newcastle upon Tyne

H. R. Rickerby Esq., Lingdale School, Yorkshire

Whitby Museum

The Trustees of the British Museum

The photograph on the Contents page is Crown copyright and is reproduced with the permission of the Controller of Her Majesty's Stationery Office.

INTRODUCTION

Nobody has seen a live dinosaur. This would only be possible if we could travel back in time for seventy millions years. But we do know what these great beasts were like from their skeletons, the popular giants of the natural history museums. These skeletons which turned to stone after long burial in the earth are fossil bones. So the world of fossils conjures up for us a vivid picture of these monster reptiles – fighting, biting and dying. And there the picture ends, for this is all that we know about fossils and the history of life on earth.

In this book we shall take a closer look at many different kinds of fossils, including those of dinosaurs. From the specimens we collect we shall learn much more about life in the past.

The study of fossils is called palaeontology, pronounced pal-ay-ont-ology. This is a word derived from Greek roots, thus: *palae*– ancient, *ontos*–form of life and *logos*–discussion; so the word simply means discussion and study of ancient forms of life. The names of the fossils themselves are made from both Greek and Latin words and there is a good reason for this. They are used by palaeontologists all over the world, and so make an international language for the science of palaeontology. These names soon become familiar when specimens are handled and labelled, and they sound very grand and distinguished, like splendid titles – ammonites, belemnites, dinosaurs, trilobites – quite fascinating!

In the past people collected fossil shells, such as ammonites, as curiosities and freaks of nature. They could not imagine that these shells were millions of years old and once belonged to living creatures of the sea. The first men to understand the importance of fossils were miners and land surveyors, who found fossil shells in the course of their daily work. They observed that the different layers of rock contained different kinds of fossils, so that each layer could be traced and recognised by its own special fossils. The rock beds were once the floors of ancient seas and lakes on which the shells had rested and afterwards been buried. From this it was seen that, in layered rocks, the oldest rocks were at the bottom and these were covered over from time to time by new or younger layers.

The rocks and fossils will show us what the world was like in the past, but only when we study them carefully and learn their secrets.

Clavilithes macrospira 80 mm
Eocene gastropod

7

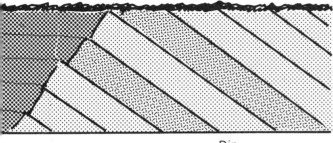
Dip

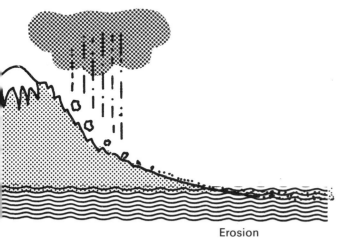
Erosion

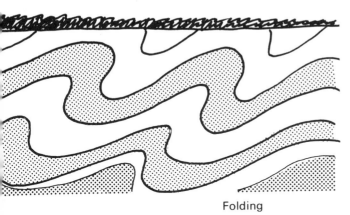
Folding

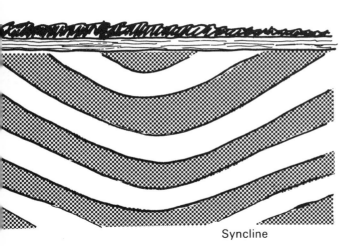
Syncline

GLOSSARY

ANNELIDA The phylum or major division of the animal kingdom which includes all worms.

ANTICLINE The form made by bedded rocks that have been pushed up in the shape of an arch or dome.

ARGILLACEOUS Rocks containing clay (Latin *argilla*–clay).

ARTHROPODA 'Jointed legs'. A phylum that includes spiders, crabs, insects and the extinct trilobites.

BASALT A hard black rock of volcanic origin. It will not contain fossils.

BRACHIOPODA 'Arm leg'. Primitive shell-fish.

BRYOZOA 'Moss animals'. Small sea animals that live in colonies. Also called POLYZOA.

CHALK A white rock formed from the limy fragments of ancient sea life. It is fossiliferous.

CHORDATA The phylum of all creatures with spinal cords–mammals, reptiles, fish and some lesser animals.

CLAY A soft plastic rock formed by erosion and sedimentation. It may contain fossils.

COELENTERA 'Hollow intestine'. The phylum of sea animals such as corals and sea anemones.

DIP The angle or slope of bedded rock.

ECHINODERMA 'Spiny skins'. The phylum of marine life including sea urchins, starfish, and the extinct blastoids and cystoids.

EROSION Wearing away of rocks and land surfaces by ice, rain and wind.

FAULT A crack in the rock crust of the earth.

FLAGS Flat slabs of sandstone that split easily along the line of bedding.

FLINT A very hard rock found as lumps in chalk.

FOLDING Ancient layered rocks, that once lay straight and level, pushed into folds by earth movement.

GRANITE A hard rock with large crystals, formed by great heat and slow cooling. It will not contain fossils.

IGNEOUS 'Formed by fire'. Rocks that were once hot soft masses and have cooled and hardened. Basalt is an igneous rock.

LIMESTONE Rock which is made up of lime or calcium carbonate. It was laid as a sea bed and will have marine fossils.

MARBLE A metamorphic or changed form of limestone.

METAMORPHIC 'Changed form.' As dough is baked to become bread, so sedimentary rocks may be 'baked' or changed by heat and pressure in the earth. Marble and slate are metamorphic rocks.

MOLLUSCA 'Soft bodied'. The phylum of all shellfish except brachiopods, but including the extinct ammonites and belemnites.

OUTCROP A rock surface clearly exposed at land level.

PORIFERA 'Having pores'. The phylum of sponges.

PROTOZOA 'The first animals'. The phylum of single cell animal life. The foraminifera are protozoans.

PYRITES Iron sulphide, a mineral of bright metallic appearance ('fool's gold') and found as mineral replacement in fossils.

QUARTZ Crystals of silica, the chief substance of the earth's crust.

SANDSTONE Rock made by sedimentation from sand grains. It is usually fossiliferous.

SEDIMENTARY Rocks formed from sediments or deposits in lakes and seas. Sandstone, limestone and shale are sedimentary rocks.

SHALE An argillaceous rock that splits easily in thin sheets on the bedding plane. It contains fossils.

SILICA As flour is the main ingredient in a pie crust, so silica (silicon and oxygen) is the chief substance of the earth's crust. Granite and sandstone are silicate rocks.

SLATE A metamorphic rock made from shale.

SYNCLINE A downward curve or dip in bedded rock, forming a bowl-shaped depression.

WEATHERING The wearing action of the weather – rain, ice and wind – on a rock face. Fossils are revealed on the surface in this way.

Anticline in limestone

RECOGNISING FOSSILS

Fossils may be difficult to find when they are embedded in the rock surface. They are usually the same colour as the surrounding rock or matrix, and only their special shape or outline may reveal them to the keen eyes of the collector. Look carefully at these six photographs. Do you recognise all of them as fossils?

This specimen has been prepared and polished. The white tubes revealed in the polished surface of the stone are corals, the skeletons of small sea creatures that lived 300 million years ago.

Lithostrotion species
Carboniferous coral — actual size

Neuropteris species
Carboniferous fern — actual size

The leaf shapes set on a central stem make this recognisable as a plant fossil. It is a fern that once grew in a forest swamp in Carboniferous times (see p. 28) and so it is about 300 million years old.

This shell was taken easily and cleanly from the soft sandy parent rock. It is the youngest fossil of this group of six, for it is only one million years old. Except for one important difference, it is like an ordinary whelk shell. Compare it with any other coiled shell, either living or fossil, and you will see that the spiral, or whorl, turns in a contrary direction. So it is named *Neptunea contraria*.

Neptunea contraria
Pleistocene whelk — actual size

Both the coral and the shell are fossils of animals with soft, boneless bodies and are called *invertebrates*, that is animals without backbones. Here, in this first picture, however, is a fish, where the skeletal form is clearly seen, with the backbone or vertebral column supporting the head, ribs and tail. Fish were the first vertebrates. The fossil comes from Solenhofen in Germany, where fine grained limestone was quarried for use in lithographic printing. Long ago this fish swam in the sea, and now the fossil is found on dry land.

Leptolepsis spattiformis
Jurassic fish – actual size

Study this rock surface. You will see the outline of three toes, pointed at the ends. This is a fossil footprint, made by a three-toed dinosaur when it walked in the mud 135 million years ago. The mud hardened to form a mould which was filled in with light coloured sand. This sand became sandstone and is the rock you now see. From the size of the print we can guess that the creature was about as big as an ostrich. Footprints and tracks are important fossils.

Dinosaur footprint – half size
Jurassic reptile

To most people this is just a large rock of no special interest. However, the group of geologists who found it on a beach recognised it as an important fossil. Its story is told on page 58. Fossil remains may be whole skeletons or separate bones and can be found complete, such as the insects trapped in amber seen on page 44. Worm borings, seen as holes in the rock, are evidence of ancient life, and so they too are fossils.

An important fossil?

MARINE FOSSILS

Oyster bed and fossil driftwood in shale
Jurassic

Most of the fossils you collect will be small sea animals, such as corals, starfish, sea urchins and shellfish. These are plentiful because conditions in the sea favoured their preservation. As a blanket is laid on a bed, so layers of sand gradually settled in the water, covering the dead animals as they lay on the sea bed. The layers of rocks formed in this way are rightly called beds.

EROSION AND SEDIMENTATION

Glaciers carve and scoop out the mountain rocks, and wind, rain and frost wear down the high lands. The work of these natural forces is called EROSION, and the broken rock material is borne away by river and flood to the sea. Here the pebbles and sand, carried as sediment in the water, settle down on the sea floor. Rock beds are slowly built up by this process of SEDIMENTATION, to become SANDSTONE, which is known as a SEDIMENTARY rock. Other sedimentary rocks, CHALK and LIMESTONE, are formed by sediments made from the chalky remains of tiny sea creatures. Fine mud may be deposited as CLAY or SHALE in deep water, or in slow moving rivers and deltas.

All these rocks were formed under water and they contain many kinds of fossils. On this and the opposite page are examples of sedimentary rocks with fossils. Collect specimens of these rocks and learn to recognise them. Some feel rough to the touch and others are smooth and greasy. Some shales are black in colour and chalk is white or cream. Test the rocks for hardness by scratching them with your fingernail; sandstone will wear the nail down, while a soft rock will show the scratch mark.

12

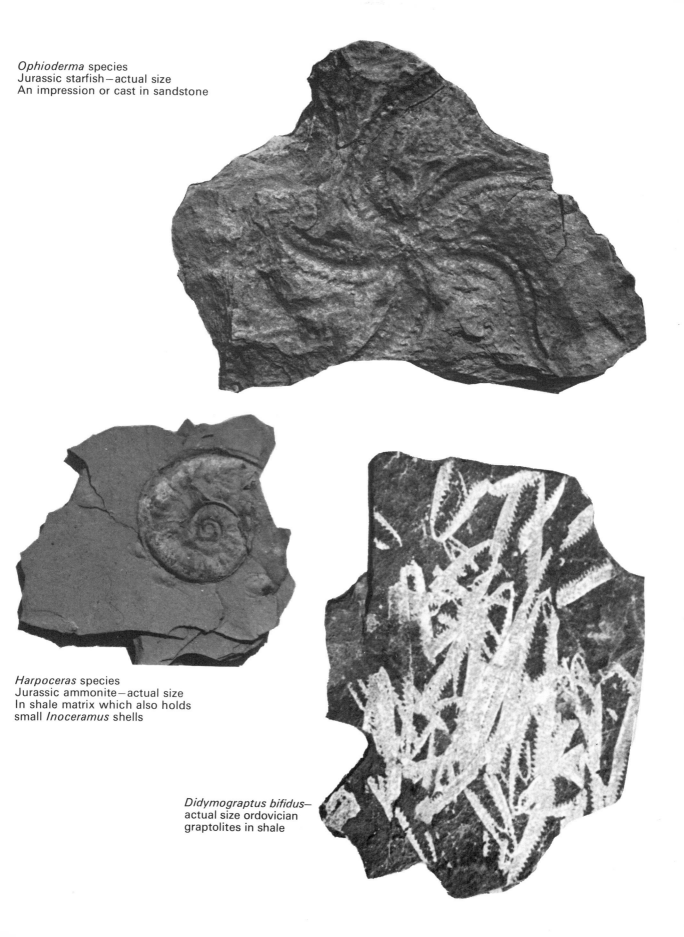

Ophioderma species
Jurassic starfish—actual size
An impression or cast in sandstone

Harpoceras species
Jurassic ammonite—actual size
In shale matrix which also holds
small *Inoceramus* shells

Didymograptus bifidus—
actual size ordovician
graptolites in shale

FIELD WORK

Fun with palaeontology starts when you lay this book aside and make your first trip in search of fossils. You learn more about them in a day's work among the rocks than from any number of books. But you do need to know where to look for fossils and be able to identify them. So plan your fossil hunt beforehand. Study the geological maps and handbooks of the area at the local library or museum, and make notes in your field book and mark places on your own map.

Fossils are found in sedimentary rocks exposed at cliffs and gullies, road and railway cuttings, quarries and mine heaps. You may need to obtain permission to enter quarries and private land.

EQUIPMENT

A sling bag or rucksack; a heavy geological hammer and a cold chisel; map, compass and a magnifying glass or pocket lens; eye shield or sun glasses to protect your eyes against rock splinters; a penknife; steel point scriber; newspaper and cartons or polythene bags for your specimens; tie-on and stick-on labels to name and number them.

Set out a page of your field book like this:

LOCATION	MAP REFERENCE	DATE
SYSTEM	BEDS	
SPECIMENS	EXTRACTED FROM	
1		
2		
3		
4		

Record your finds carefully, for unless a fossil is known to come from a certain place and rock bed it is scientifically useless.

At the location look for fossils that are exposed on the rock surface and decide on how they can be taken out without damage. Extracting fossils needs care and patience, but you will soon learn the knack. Soft rock and delicate shells call for a light touch, and it may be necessary to cut out a sizable piece of rock with the fossil to avoid breaking it. Some fossils set in hard rock jump out cleanly with a smart blow of the hammer struck at the rock surface. Record each good specimen, label it and wrap it in newspaper. Discard poor specimens.

Coroniceras species 65 mm
Jurassic ammonite

14

FIELD EQUIPMENT

Rucksack
Hammer
Cold chisel
Pen knife
Steel point scriber
Pocket lens

Field book
Maps
Newspaper
Polythene specimen bags
Sun glasses

Pleurotomaria anglica 55 mm
Jurassic gastropod before and after
cleaning

PREPARING SPECIMENS

A preparation board for cleaning and developing specimens can be made from a few pieces of timber. Use a square of block board as the base, making it about 25 cms square. Cut out two rails, 2×4 cms, and screw these to the base to make a bench hook. Two wooden turn-buttons are screwed to the top rail, and the board is made.

Flat slabs of rock can be held by the turn-buttons, and a G clamp is used to hold large, irregular pieces. With the specimen firmly held on the board, the work of cleaning begins. Use needles, penknife and small jewellers chisels to chip away the matrix rock. Take off small flakes, at right angles to the fossil, and so avoid scratching the surface. Small fragile shells and plants are left on the slab, after being cleaned and exposed. To release a fossil from the matrix, it may be necessary to drill holes around it, and then cut it clear. As fresh cut rock hardens quickly in the open air, rough trimming should be done as soon as possible after the fossil is taken.

Weak acids are used to dissolve and loosen limy rocks, and acetic acid or vinegar are suitable for this purpose. The *Pleurotomaria* specimen was cleaned by first chipping away all the matrix, and then coating the exposed shell with a film of wax before brushing on dilute hydrochloric acid (5% dilution) to remove the rock in the crevices. This waxing prevents the further action of the acid on the specimen. Finally the fossil was scrubbed in hot soapy water with an old toothbrush. Incidentally, a word of warning should be given about hydrochloric acid, as this can be dangerous unless extreme care is taken to avoid splashing. If you do propose to use this acid, the solution must be dilute, and it would be advisable for an adult to prepare it.

Fossil shells taken cleanly from soft beds do not need much preparation beyond dusting off loose dirt. Before you do this, examine the specimen with a hand lens, for you may find a growth of *bryozoa*, or the cast of a *serpula* worm. One of the shells on the opposite page was deliberately left unclean for this reason.

Broken fossils can be repaired with synthetic glue. After preparation the specimens should be labelled and displayed in their various scientific classes. (See p. 36).

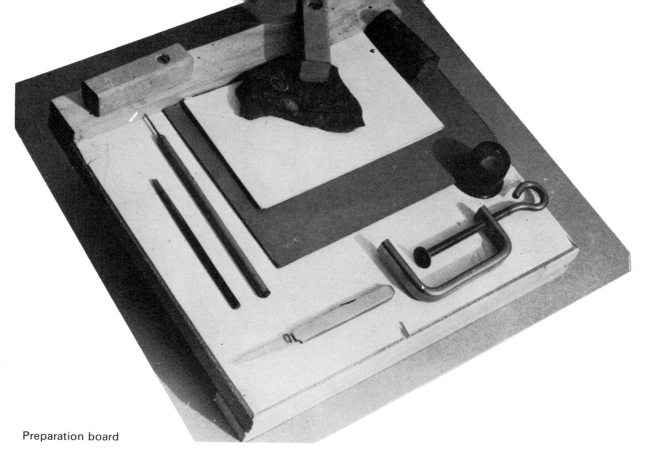

Preparation board

Labelling specimens

17

The British Junior Naturalists Association on a fossil hunt

THE FOSSIL CLUB

These young people are members of a naturalist association, and their outing was planned so that all could learn how to find and identify fossils and share the interest of the hunt.

There are many advantages in belonging to a club or society, which can, for instance, be formed on a school or neighbourhood basis. Here are some rules you should follow:

Plan your field work beforehand and obtain as much information as possible about the area of your visit.

Wear suitable clothes and carry food and equipment in a sling bag or rucksack.

Observe a code of conduct; do not leave litter; obtain permission to enter private ground; do not expose yourself or others to danger—at the sea coast watch the tides and beware of dangerous cliffs; protect wild-life. In remote country leave information of your route with someone at your base or home.

Report important finds to your museum.

Museum examples of the fossils
you seek

MUSEUM STUDY

Your own field work takes first place in the study of palaeontology, and museum study might come next, for here you will see examples of the fossils you seek. The museums of the large cities have many rare specimens on exhibition, such as the dinosaurs and sea reptiles. Many hours of patient work go into cleaning and mounting these exhibits.

Most town museums will have a geological section where local fossils are displayed, and these may be of special interest to you. Get to know them well and become an expert on the rocks and fossils of your own district. This is the beginning of serious study in palaeontology. Learn to classify and label fossils by your own research, using reference books and museum exhibits for this purpose. The museum staff will not be greatly interested in fossils that are commonly found and already well represented in the collection, and you should seek their help only when you are certain that you have a rare or exceptionally good specimen. The museum will have a collection of geological maps and handbooks which you should consult, and the staff will give you information about fossiliferous localities and local societies you might join for field work.

Young palaeontologists will enjoy visiting the national museums, such as those listed below. They have splendid fossil displays as well as literature for sale and programmes of lectures and films.

H.M. Geological Survey and Museum,
South Kensington, London SW7

British Museum (Natural History),
South Kensington, London SW7

Royal Scottish Museum,
Chambers Street, Edinburgh 1

Typical Limestone Joints

FIELD SKETCHING

A geological map of your district is different from a surface map, whose features you may know quite well. Your own sketches will help you to follow the arrangement of rocks that are shown on the geological map if you use the special signs that represent the rocks; sandstone is shown by dots, limestone by open lines with joint marks and clay rocks by close lines.

The sketch opposite is of a river bank, with a low cliff exposure of limestone. Date, place and map reference are written at the top and also the name of the 'country' rock, or local formation. Two curves are drawn to show the top soil and a few rough strokes indicate the trunks of fir trees. A lower line marks the depth of a clay deposit, and under this is the limestone formation, marked with joints. The drawing carefully shows where the beds dip at an angle to the North. Fossil finds are marked with a cross and numbered to correspond to specimens taken. The river level is drawn with a few wavy lines. Also shown is the grading action of the river current, when the larger rocks are caught on the spur of the bend and the smaller pebbles are carried further downstream.

You will understand the need for notes and sketches when you return home with a bag full of fossils, for now you have to sort them, and recall where you found them. It should again be repeated that the specimens are scientifically useless unless you know exactly where each one came from.

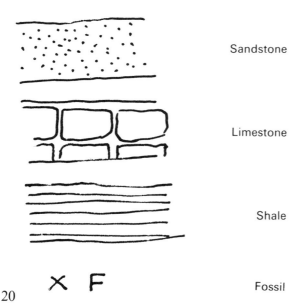

Sandstone

Limestone

Shale

Fossil

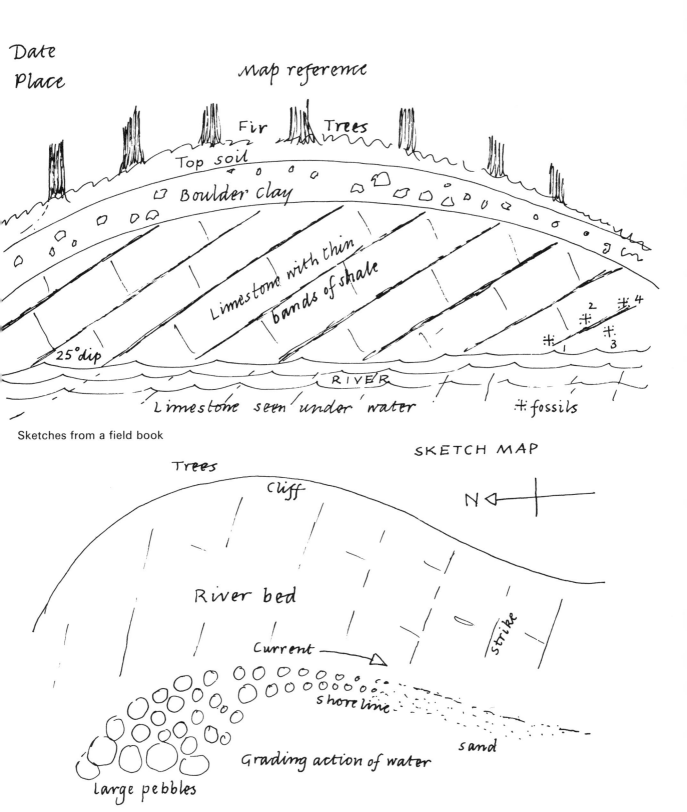

Date
Place

Map reference

Fir Trees

Top soil

Boulder clay

Limestone with thin bands of shale

25° dip

RIVER

Limestone seen under water

⚹ fossils

Sketches from a field book

SKETCH MAP

Trees

Cliff

N ◁

River bed

Current ⟶ ▷

strike

shore line

sand

Grading action of water

large pebbles

21

PHOTOGRAPHY

You should make good use of your camera in the study of fossils. The fossiliferous rocks can be photographed when you are out collecting, and you can make prints and slides of your specimens at home.

Record the places you visit, which may, for instance, be a beach or a quarry, by first taking a long shot or general view of the landscape. Next take a medium shot at a few metres distance, to show the nature and structure of the rock faces. The fossils can then be photographed at close range before extraction. These close-ups are important because you may not, in fact, be able to extract the fossil. The large ammonite on the opposite page was not taken because it was impossible to remove it in one piece, and it was also a sea worn specimen. Notice that the hammer is placed alongside it to give an idea of the size. The hammer is 40 cms long, so this makes the ammonite about 50 cms in diameter.

Fossils also make good subjects for home photography. This may be done in daylight or artificial light, with the subject well lit on all sides to avoid heavy black shadows. Use white paper or card to throw reflected light into shaded areas. You should take some care in focussing and lighting, so that you get sharp detailed negatives. From these your prints should be enlarged to the same size as the fossil, when the clear detail will be shown.

Colour positive film gives you the opportunity to make a set of slides, which you could then use to illustrate a lecture on fossils. Select good specimens and shoot them against a light neutral background, to give value to the tones of the fossils.

Cardinia species 85 mm
Print your specimens same size

General view of a locality—long shot

Close up showing a large ammonite

A DIORAMA

Models of the dinosaurs described on page 53 can be bought as kits, to be made up, or you can model your own in clay as explained below. You can also make simple flat models by drawing the reptiles on white card and cutting them out. Display your models in a diorama. This is an open front box like a theatre stage, where the dinosaurs can be placed against a scenic background. The model described here is collapsible, and can be laid flat and stored away tidily.

Make the diorama from light strong cardboard of the kind used for large cartons. Cut four pieces to size as shown in the diagram and hinge them together, using wide adhesive tape. For the front make a wooden frame from material 4×2 cms thickness. The joints can be dovetailed, or more simply nailed and braced with angle plates. Attach the frame to the box with thumb tacks. A length of thin card is set in the box to form a continuous curved background. This hides the back corners, and also gives the illusion of space when coloured light blue to represent the sky. A piece of glass or mirror placed on the base and surrounded with pebbles and sand will suggest a pool or a shore. As there were no flowering shrubs and plants until the end of the time of the dinosaurs (the Cretaceous), be sure that the vegetation is in keeping with the period. Use moss and small ferns to dress the set; clubmoss and pieces of that troublesome weed *Equisetum arvense*–field horsetail–are suitable and can be set in small knobs of clay.

Using the drawing on page 55 as a guide, you can make a clay model of the dinosaur *Triceratops*.

1. Take a length of soft, thick wire and form a loop to the shape of the head, then bend the wire to the curves of the spine and tail.
2. To make the forelegs twist on a length of wire at the shoulder, leaving two ends to be bent to the shape of the leg bones. Form the hind legs in the same way at the hips.
3. Wrap string around the wire skeleton; this will hold the clay as you begin to model. Build up the model gradually with small pieces of clay to the general outline of the drawing.

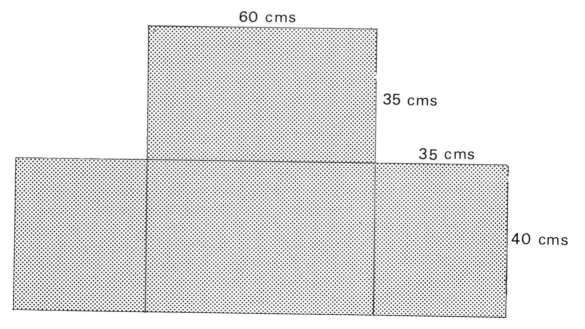

60 cms

35 cms

35 cms

40 cms

Box and frame assembled

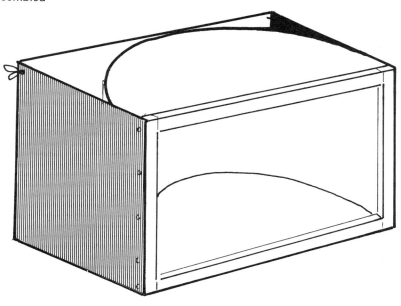

Models in a diorama
Left: *Brontosaurus* Right: *Stegosaurus*

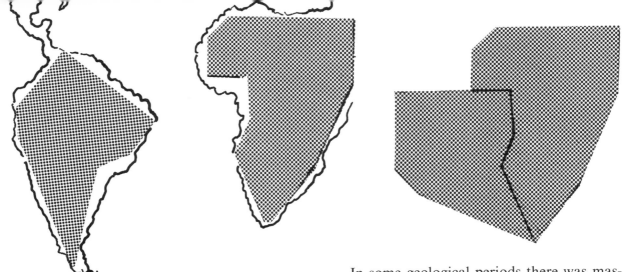

ANCIENT LANDS AND SEAS

Trace the shaded parts of this map on thin card and cut them out. They will fit neatly together like pieces of a jigsaw puzzle. Geologists believe that Africa and South America were once joined in a large continent, together with India, Antarctica and Australia. This land mass began to split up in Palaeozoic times and the parts then slowly drifted asunder. The lost continent is called Gondwanaland – from Gond, a region of ancient rocks in India.

In some geological periods there was massive disturbance of the earth's crust. Mountain ranges were formed by upheaval and folding of the rock STRATA, or layers of rock. Scientists call this mountain building OROGENY, from the Greek *oros* – mountain, and *genesis* – birth. The mountain chain formed by the Rocky Mountains and the Andes is an example of giant orogenic forces at work. The Alps and the Himalayas are considered young mountains in geological time, and the rocks which form them were once the bed of an ancient ocean, the Sea of Tethys, so named after the Greek goddess of the sea. It is therefore not surprising that fossil sea shells are found on Mount Everest.

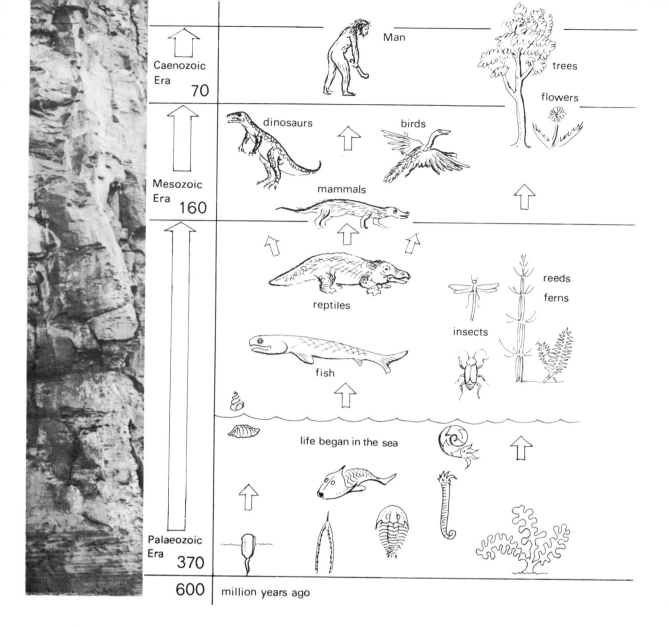

			Man	trees
Caenozoic Era **70**				flowers
Mesozoic Era **160**	dinosaurs	birds		
	mammals			
	reptiles	insects	reeds	ferns
	fish			
	life began in the sea			
Palaeozoic Era **370**				
600	million years ago			

THE FOSSIL RECORD

The record of the rocks and their fossils is divided into three great ERAS of time:

1. THE PALAEOZOIC ERA – the age of ancient life
2. THE MESOZOIC ERA – the middle age
3. THE CAENOZOIC ERA – the age of recent life

Each era is divided into PERIODS of time which are recorded in SYSTEMS of rocks. The rock systems are then divided into STAGES and ZONES. These may have local names; for example, the *Lias* in England and the *Morrison Beds* in the U.S.A. are both stages of the Jurassic System.

The fossils show a gradual progress of life – from creatures of the sea to mammals and Man.

Cephalaspis species
Palaeozoic fish—actual size model

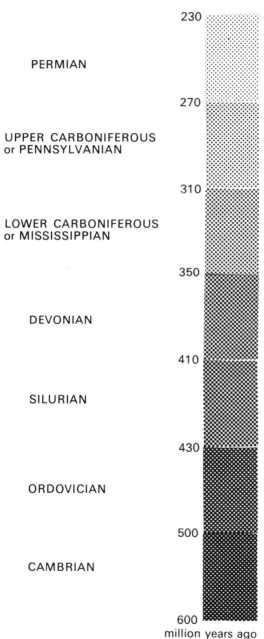

230
PERMIAN
270
UPPER CARBONIFEROUS
or PENNSYLVANIAN
310
LOWER CARBONIFEROUS
or MISSISSIPPIAN
350
DEVONIAN
410
SILURIAN
430
ORDOVICIAN
500
CAMBRIAN
600
million years ago

THE PALAEOZOIC SYSTEMS

The geological column shows the systems of rocks which make up the Palaeozoic Era. Each named system of rocks also represents a period of time, measured in millions of years. Fossil history begins with the Cambrian period, seen at the bottom of the column, as the lowest system of rocks.

The nineteenth century geologists named these old rocks when they first studied them in Wales. Cambria is the Roman name for Wales, and the Ordovices and the Silures were the ancient people of Cambria.

Life began in water, and all Cambrian fossils are of sea animals, that is TRILOBITES (opposite), shellfish and seaweeds. Marine life continued in the Ordovician period, but there is no evidence in the fossil record of life on land. BRACHIOPODS (primitive shellfish—see page 46) and GRAPTOLITES (opposite) are common fossils of this time. The first CEPHALOPODS (ancestors of the octopus—see page 50) also appeared then. The land was still barren waste until the Silurian period, when land plants first became established. The oxygen released into the atmosphere by green plants was needed to sustain later animal life on land. Primitive types of fish, such as *Cephalaspis*, made their appearance at this period as the first vertebrates.

Devonian rocks are well exposed in the county of Devon, from which they take their name. The period is called the Age of Fishes,

Didymograptus species 30 mm
Ordovician graptolite—tuning fork type

which were then well established in the seas.

Carboniferous simply means 'lots of coal' and the coalfields of today are the fossilised remains of the great swamp forests of Carboniferous times.

Permian rocks take their name from the district of Perm in the U.S.S.R. The Permian period marks the end of ancient forms of life on earth, the closing of the Palaeozoic Era.

TRILOBITES These attractive fossils prove that a well developed form of animal life flourished in the ancient seas, but all the trilobites are now extinct. They were ARTHROPODS, like crabs and lobsters, with the legs and body protected by a hard shell. You see that the body is shaped in three parts or lobes, and this is why the trilobite is so called—TRI-LOB-ITE.

The fossil *Calymene* is often found rolled up, in the same way as the wood louse closes into a ball for defence. *Agnostus* was a small, blind trilobite that crawled on the sea bed, but many of the larger trilobites were good swimmers and had large compound eyes.

Label the specimens in your collection—

Class: TRILOBITA

Genus: . . . .

Species: . . . . (see p. 36)

GRAPTOLITES These little sea creatures are extinct, for they completely died out at the end of the Silurian period. They were tiny polyps, little tubes of jelly, that lived together in a shell tower, each with a separate cell. The fossil shells are found as branches or polyparies.

Graptolites and trilobites are important to the geologist as guides to the Palaeozoic rocks of the world.

Calymene blumenbachi 50 mm
Silurian trilobite

Olenus gibbosus 20 mm
Cambrian trilobite

Agnostus pisiformis 8 mm
Cambrian trilobite

29

Pholidophorus species 100 mm
Jurassic fish

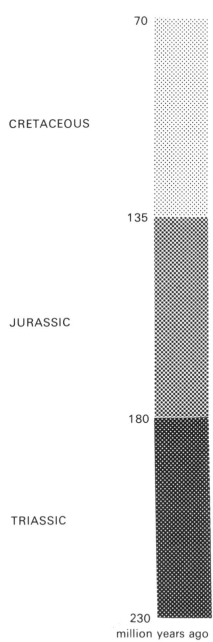

70

CRETACEOUS

135

JURASSIC

180

TRIASSIC

230
million years ago

THE MESOZOIC SYSTEMS

The middle era of life history, the Mesozoic, may be called the Classic Age of Fossils, for these are many and varied in the Triassic, Jurassic and Cretaceous rocks of the world. Reptiles were the dominant animals on land, and giant turtles and amphibians ruled the sea. The huge reptiles known as DINOSAURS lived in this age.

Triassic rocks were first studied in Germany, where they were classified into three groups and named the Trias, which means three groups of rocks. The nature and colour of the rocks suggest that much of the land was dry, hot desert, which would be suitable for reptilian life. Fossil fragments have been found of the earliest mammals. They were small rat-like creatures.

The Jurassic system is named after the Jura Mountains of France. AMMONITES and BELEMNITES (see p. 50) flourished in the warm Jurassic seas, along with seagoing reptiles and many kinds of fish. *Pholidophorus* is a good example of the streamlined, bony fish of this time.

Cretaceous means 'chalky', from the Latin word *creta*–chalk. Much of the rock laid down at this time consisted of the limy fragments of sea plants and animals that sank in quiet waters and gradually built up to chalky rock. The period saw mammals established on land, but reptiles were on the decline and the great race of dinosaurs died out completely.

Pleurotomaria anglica 55 mm
Jurassic gastropod

The shape of *Rhynconella* is like a nose, and that is what its name means. It is a primitive shellfish or brachiopod. The other fossils are molluscs (soft bodied), like our modern shellfish. *Pleurotomaria*, with a coiled and ornamented shell, was a sea snail, and *Gryphea* was an oyster, with two unequal shells, or valves. The larger righthand valve curls in to form the umbo or hinge point, where it meets the smaller left valve. *Lima* is a large, smooth bivalve, and it is seen next to the mould or impression it made in the matrix rock. All these are common fossils that are collected in a Jurassic area.

Gryphea arcuata 70 mm
Jurassic oyster

Rhynconella tetrahedra 30 mm
Jurassic brachiopod

Lima gigantea 180 mm
Jurassic lamellibranch

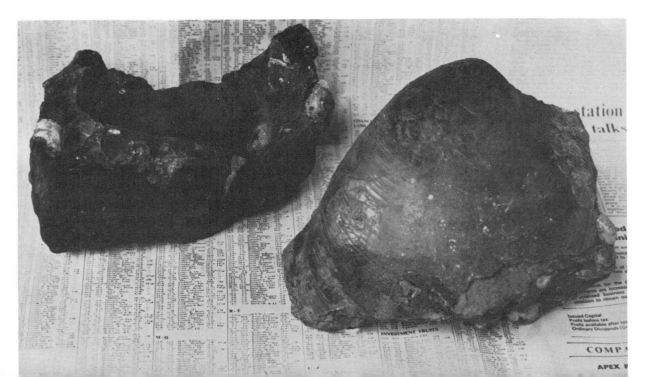

Eohippus species 30 cms
Eocene mammal

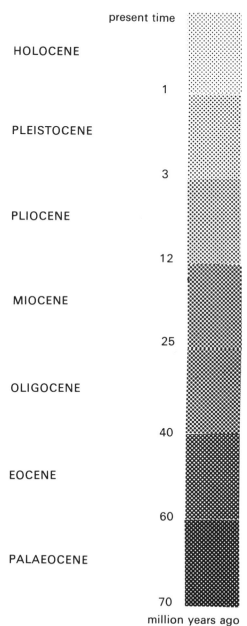

present time

HOLOCENE

1

PLEISTOCENE

3

PLIOCENE

12

MIOCENE

25

OLIGOCENE

40

EOCENE

60

PALAEOCENE

70

million years ago

THE CAENOZOIC SYSTEMS

The Caenozoic is the shortest of the three eras, as it covers only the last 70 million years of life history. It is called the Age of Mammals, when these warm-blooded animals rose in great numbers to replace the reptiles as lords of the earth.

The names of all the systems contain the Greek root CEN, which means recent. So, first of all, we have the Palaeocene–the ancient recent period, the Eocene–the dawn of the recent and the Oligocene–the least recent. These three systems may be considered together. All the fossils of this time show the rapid advance of animals and plants towards the types of our modern world. The earliest ancestor of the horse is found as fossil bones in North America. It was a small mammal thirty centimetres high, and is called *Eohippus* –the dawn horse. Huge size is no guarantee of survival, for dinosaurs died away completely and the smaller mammals increased. They adapted themselves to climate and conditions of life in the forest and on the grasslands. Unlike the slow moving reptiles, they became nimble and alert. Grass eating animals developed long limbs and hoofed feet for running; tree dwellers used their forelegs as arms for climbing. This was the time of the first monkeys.

In the Miocene (less recent) and the Pliocene (more recent) periods all the modern

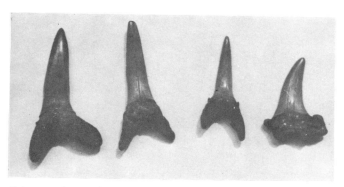

Odontaspis species 25 mm
Eocene shark teeth

Chama squamosa 40 mm
Eocene oyster

Turricula rostrata 70 mm
Eocene gastropod

Sycostoma pyrus 55 mm
Eocene gastropod

mammals had appeared, except Man.

It was in deposits of the Pleistocene (most recent) period that bones of ape-like men were unearthed. During the Pleistocene there were four great invasions of ice from the North Polar Cap, which covered much of Europe and North America. The time of the Ice Ages ended ten thousand years ago, at the beginning of the final period of fossil history. With the Holocene (wholly recent) the fossil record ends.

Sharks' teeth are found in plenty as fossils of the early Caenozoic systems. Here they are seen, fresh looking, sharp and polished, when all traces of the soft bones and flesh of the shark have gone. The teeth are covered with hard enamel, and this accounts for their excellent preservation over a period of 20 million years.

Chama squamosa is an Eocene oyster which should be compared with the Jurassic *Gryphea*. *Turricula rostrata* and *Sycostoma pyrus* are sea snails, or GASTROPODS, that lived in warm shallow waters of Eocene time, when the world climate was much hotter than it is now.

Harpoceras species 20 mm
Jurassic ammonite pyritised

MINERAL REPLACEMENT

This ammonite was found bright and shining in the dark shale, as though made of gold. The original shell was replaced by iron pyrites – 'fool's gold'. Water in the rocks had gradually dissolved the shell, and at the same time deposited the iron salt, in a swopping action called molecular replacement. The bright metal of pyritized fossils may be preserved by a coating of acetate varnish.

Limy shells may be completely dissolved in the rock, leaving only a mould or impression. The *Pseudopecten* fossil is a hollow mould made by a scallop shell.

Pseudopecten equivalvis 140 mm
Jurassic bivalve

FOSSILS AND STRATIGRAPHY

Stratigraphy is the study of STRATA, the layers of rock that form the earth's crust. The diagram shows a rock face of the Jurassic system, with the strata marked in stages and zones. Each zone is known and named by a special fossil found in it, and belonging only to that zone.

1. Zone of *Lytoceras jurense*
2. Zone of *Hildoceras bifrons*
3. Zone of *Harpoceras falcifer*

These are different species of ammonites (see page 50), and ammonites make good zonal fossils because they are found abundantly in Jurassic rocks all over the world. On the left of the diagram the three zones stop abruptly at a line, and then continue lower down. This line is a crack or FAULT, where the strata have been broken by some disturbance of the earth's crust. The three zones once formed a level and continuous band of rock, for the zonal fossils are to be found in their correct order on each side of the fault. By means of zonal fossils, rocks can be recognised and matched.

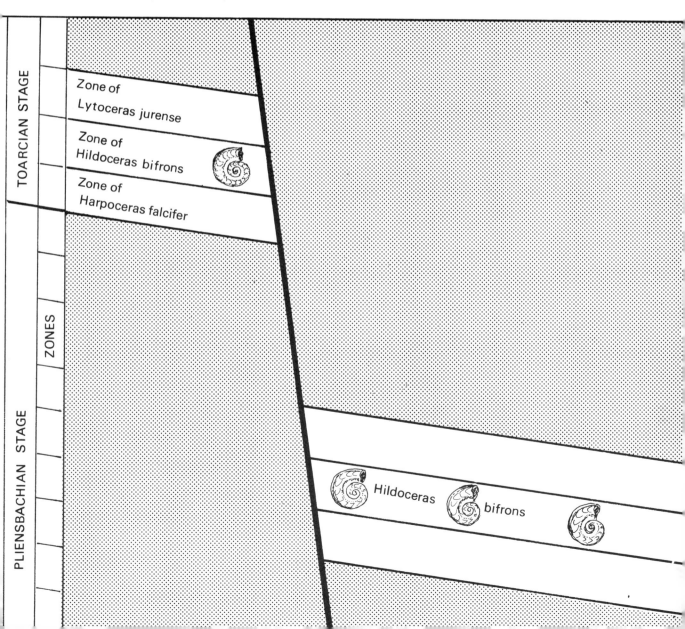

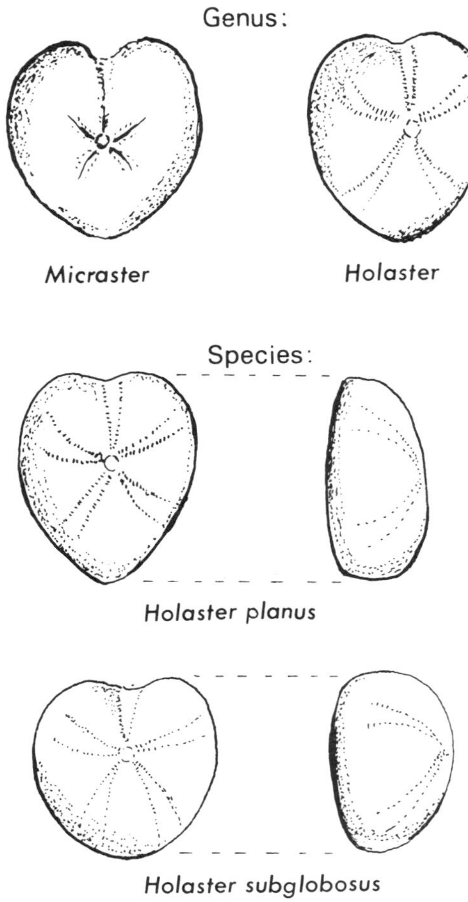

Genus:

Micraster

Holaster

Species:

Holaster planus

Holaster subglobosus

BINOMINAL CLASSIFICATION

This means that living things and fossils have a double name or binomen. The first is a general name, the GENUS, and the second is a special name, the SPECIES. These are expressed in Greek or Latin. The two fossil sea urchins shown here can both be described as being heart-shaped, but one is marked with a small star, and the other with a large star. *Aster* is the Greek word for star, so the first is of the type or genus *Micraster*, meaning *Little Star*, and the second is of the genus *Holaster*, *Whole* or *Big Star*.

Next are two species of *Holaster*, one flat in shape and the other rounded. Their binomens are *Holaster planus (Big star flattened)* and *Holaster subglobosus (Big star globular)*.

These fossils are related to the common sea urchin in the class ECHINOIDS. Unlike the latter, however, they are not a regular round shape, so they belong to the order IRREGULAR. Tracing the line down, we find that they are free swimming types of the phylum ECHINODERMS (spiny skins – see page 48).

Classification of natural life produces a Family Tree for each kind of animal and plant, and when fossils are included, the line can be

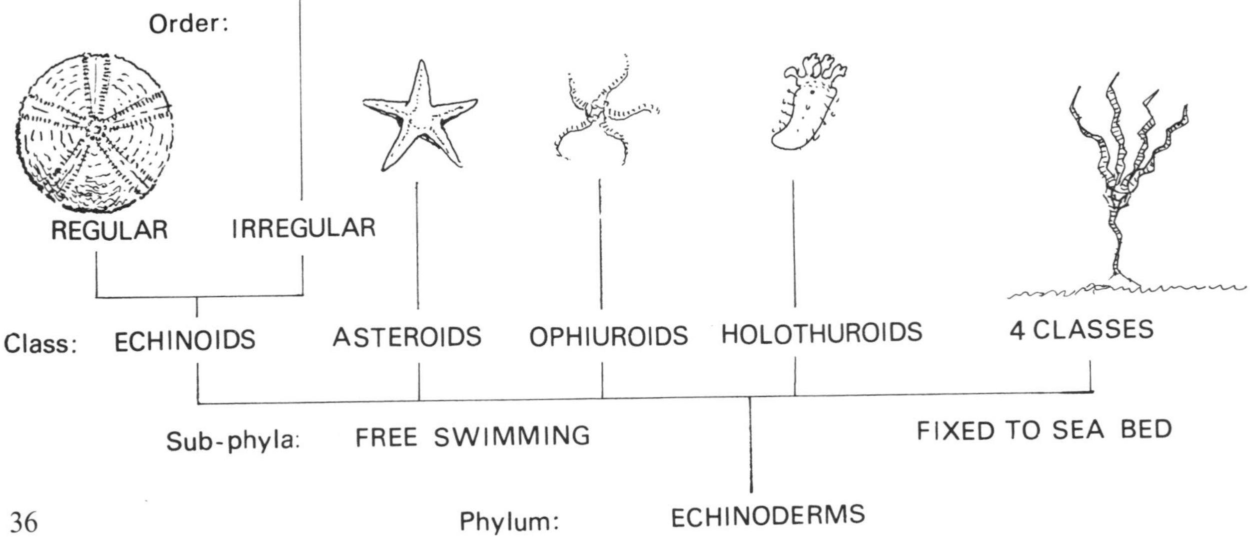

Order:

REGULAR   IRREGULAR

Class: ECHINOIDS   ASTEROIDS   OPHIUROIDS   HOLOTHUROIDS   4 CLASSES

Sub-phyla: FREE SWIMMING   FIXED TO SEA BED

Phylum: ECHINODERMS

traced to show the Origin of Species, and the process of EVOLUTION.

All living things belong to either the Animal Kingdom or the Vegetable Kingdom. The kingdoms are divided into PHYLA.

Animal Kingdom:

Phylum	
CHORDATA	Mammals, birds, fish, reptiles, amphibians
ECHINODERMA	Sea urchins, starfish, sea lilies
ARTHROPODA	Insects, crabs, lobsters, spiders
ANNELIDA	Worms
MOLLUSCA	Snails, mussels, oysters, cuttlefish
BRACHIOPODA	Lingula (see page 46), lamp shells
BRYOZOA	Moss animals
COELENTERA	Jelly fish, corals
PORIFERA	Sponges
PROTOZOA	*Foraminifera, radiolaria* (see page 42)

Nearly all the phyla of the animal kingdom are marine creatures and are invertebrates, which, as we have learned, means animals without backbones. The sea is where life began, and in early Palaeozoic times fishes were the highest form of life, as they were the first back-boned creatures of the phylum Chordata. They were also the ancestors of all the land vertebrates – amphibians, reptiles, birds and mammals.

A LIVING FOSSIL

The Coelacanth, which was once considered entirely extinct, is a lobe-finned fish, with fins attached to four lobes or limbs, like rudimentary legs of a land animal. The first lobe-fins and their relatives the lung-fish used their short legs to crawl ashore and to live from time to time on the land. Amphibians and reptiles evolved from them.

Although at one time coelacanths were found only as fossils, in 1939 a large blue fish was netted and landed at the port of East London, South Africa. It was an odd-looking creature with four legs, and the fishermen who caught it had never seen such a fish before. The curator of the local museum, a Miss Latimer, saw it, and recognised it as a fish of the sub-order COELACANTHS, a fossil type that had supposedly died out completely in the Cretaceous period. So here was a living fossil! Since then several more coelacanths have been caught, and they are now classified as the genus *Latimeria*, after the lady who first identified them. *Latimeria chalumnae* is nicknamed 'Old Fourlegs'.

Latimeria chalumnae 150 cms
Living descendant of the extinct
Coelacanths

COAL

Three hundred million years ago the lowlands of the earth were vast swamps, thickly covered with green vegetation. In the hot damp air the trees and ferns grew in profusion, then died and sank in the mud. Amphibian reptiles lived in the stagnant pools and slow moving rivers of the swamps. Giant insects flew in the forests of primitive trees and lush undergrowth.

The fossil plant life of this time is now found as carbon or coal, and the Carboniferous period is named after the extensive deposits that are mined in seams in the Carboniferous rocks.

The forest trees were conifers and lycopods, the giant club mosses. *Lepidodendron* means 'scaled tree' and refers to the leaf scars like scales seen on the stems of these trees. The fossil opposite shows the well marked pattern of the scars. *Alethopteris* is one of the many forest ferns. The leaf form is accurately reproduced in a thin carbon film on the stone. The reeded stems of giant equisetums, or horse-tail ferns, are common fossils of Carboniferous rocks.

Coal has been pressed into a black, shiny, solid mass, leaving no trace of the shapes of leaves and stems, and so you will not be able to find fossils in the actual coal itself. These occur in the sediments laid above and below the seams of coal.

Alethopteris species
Carboniferous fern—actual size

Lepidodendron species—bark surface
Carboniferous tree—actual size

Coniopteris hymenophylloides
Jurassic plant-actual size

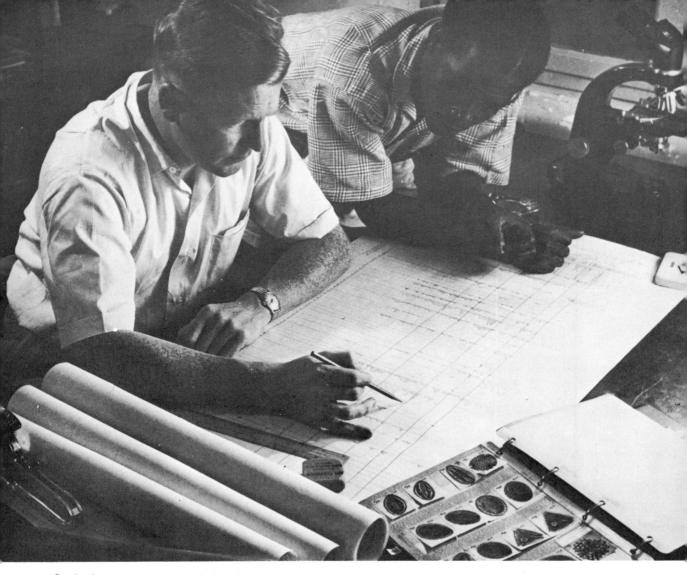

Geologists at work in search for oil
A SHELL photograph

OIL

These two men are palaeontologists engaged in the search for oil. They are making a chart of deep rock formations from samples or cores brought up from underground by a drilling rig. The samples are examined under a microscope to reveal *Foraminifera*, fossil shells no bigger than grains of dust. The foraminiferans are tiny sea animals with many different shell forms, and these zonal fossils are a very important guide to the oil-bearing rocks. The various layers of rocks are known by the fossils contained in the specimens found in the cores.

Crude oil is called petroleum, a word made up from the Latin *petra* and *oleum*, rock and oil. The oil is the remains of abundant natural life of ancient sea beds, transformed in the deep rocks into a thick fluid. The oilfield is a reservoir where the oil is trapped in an ANTICLINE, a geological formation of rocks like an inverted saucer. Oil seeps through sandstone and water-bearing rock, and then accumulates under a cap of impermeable rock such as shale or clay, where it lies trapped as an underground lake.

A simple experiment will show how oil seeps upward through permeable rock, to lie on top of rock and water. Pour old engine oil into a glass jar to a depth of one cm. Then add the same volumes of sand and water. Allow the mixture to settle, and the oil will be seen to rise clear of the sand and water.

A simple experiment

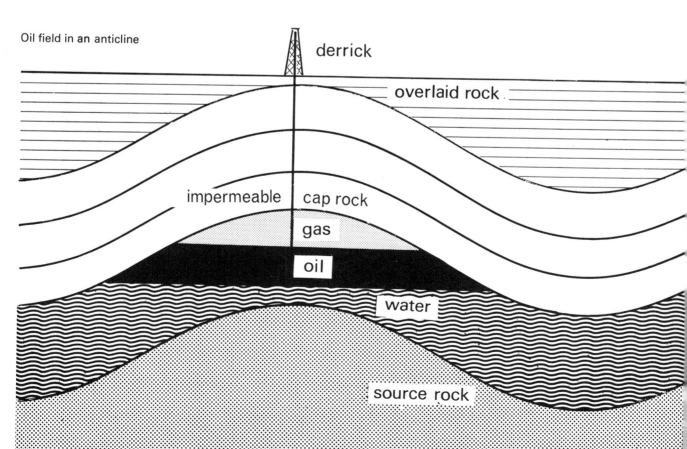

Oil field in an anticline

derrick

overlaid rock

impermeable | cap rock

gas

oil

water

source rock

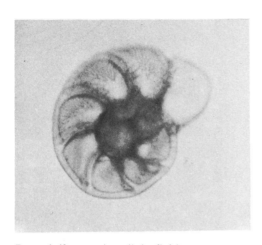

Foraminifer species—light field

MICROFOSSILS

On page 40 two palaeontologists were seen charting rock strata by means of zonal fossils. They used a microscope to examine the minute shells of the microfossils, and these are shown on the opposite page.

The beauty and astonishing detail of the little fossils is revealed by the microscope. Animals and plants can be seen in this miniature world. Most of the animals belong to the phylum PROTOZOA, and they are simple blobs of protoplasm jelly surrounding a nucleus cell. Two of the protozoans, the *Foraminifera* and

the *Radiolaria*, produce hard shells and skeletons which sink to sea floor when these marine animals die. Foraminifera are found as fossils in limestone and chalk and radiolaria occur in deep sea deposits called radiolarian ooze.

The shells of foraminifera are of lime or calcium carbonate, or in some species a covering is made from small grains of sand. Some of the shells are coiled like ammonites, and others like corn grains and peascods. The Nummulites, or coin fossils, are the giants of this Lilliputian world. They are found in Eocene rocks as disc-shaped fossils like coins of long ago.

The skeletons of radiolaria are formed of silica, which is a hard and stable mineral. They are shaped in a network of intricate tracery, like cups, vases and globes.

With their wonderful shapes and patterns, microfossils make particularly exciting subjects for photography. For the miniature camera a special fitting is used which slips over the microscope tube, the lens of the camera being removed. Good results can be achieved by using a camera with extending bellows and a ground glass focusing screen. The drawing shows how this could be set up on a board. Prepare the slides from scrapings of natural chalk and limestone made into a paste with water. Specimens already prepared and mounted may be obtained from dealers where microscope accessories are sold.

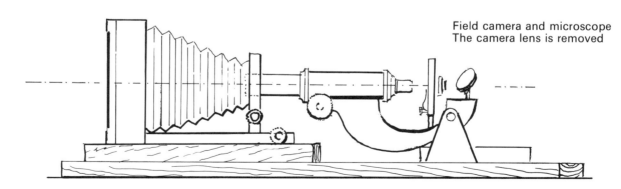

Field camera and microscope
The camera lens is removed

Foraminifera species
A SHELL photograph

Radiolaria species—dark field
Like spacecraft

43

Jet brooch 42 mm

Natural jet with impressions
of ammonites 150 mm

AMBER AND JET

Amber and jet are glassy substances of fossil origin, and both are derived from ancient trees. Jet is dense black and lustrous. It is found as pieces of fossilized driftwood in the Jurassic rocks of the Yorkshire coast at Whitby, where it has been sought and used for jewellery since the time of the ancient Britons. Beads, brooches and ornaments made of jet were very popular in the nineteenth century.

With its rich shades of red and yellow, and soft translucence, amber was also much used in the old days. It was cut and polished into beads for necklaces and bangles, and shaped and bored to make pipe stems. Polished amber pieces often revealed insects trapped inside them, and these were prized as curios. Amber was originally resin, exuded from conifer trees in the warm summers of early Caenozoic times. It was this sticky gum which became the trap for flies and small insects, where they were engulfed in the soft, oozing mass. The resin hardened and fossilized under the sea of the Baltic coast, where it is now found as amber.

Amber cigar holder 120 mm

Enlargement showing trapped
insect

Natural amber with trapped insects
95 mm

CORALS

Corals and sea anemones look like plants growing in an underwater garden, in the clear shallow seas where they are found. They are not plants, however, but animals, and are related with jelly fish and the extinct graptolites in the phylum COELENTERA. The word means 'hollow intestine', for the living coelenterate is simply a mouth and a gut formed of jelly.

Corals have a hard limy skeleton, and usually live in groups or colonies, to make massive reefs of coral rock, although some larger corals are found singly as cone shaped fossils. *Halysites*, the chain coral, is a colonial type, and there are many others. From the ancient coral reefs we can now trace the shores and coastlines of the past.

Fossil coral is found in great abundance in limestone rocks that have been formed as sediments in shallow seas. The rock is quarried and burned to make lime for farmers and builders. Massive limestone is cut and polished in large slabs and columns for buildings. The polished surface reveals the coral as a white lacy pattern against the dark matrix. Small pieces of coral limestone could be tumble polished in a pebble polishing machine, as cutting and polishing by hand is very hard work.

Halysites catenulates 90 mm
Silurian chain coral

BRYOZOA

The name means 'moss animal', and this describes the spreading moss-like habit of these sea creatures. They are also called POLYZOA, meaning 'many animals', because the skeletal mass is a colony of tiny animals, each in a separate cell. The fossil *Fenestella* takes its name from the 'windows' formed by the empty skeleton. Bryozoa are often found on the surface of fossil marine shells.

Fenestella species 100 mm
Carboniferous bryozoan

Enlargement showing fenestration
or windows

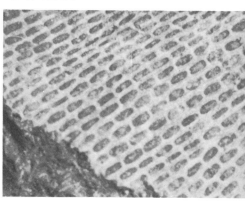

BRACHIOPODS

BRACHIOPODS are shellfish of a primitive type, the living creature being anchored to the sea floor by a stalk or pedicle which emerges from the beak of the shell. Common as fossils of the Palaeozoic and Mesozoic eras, they are now a declining race. It is important to learn the difference between brachiopods and bivalves of the phylum MOLLUSCA, such as scallops. Notice in this drawing that the brachiopod is

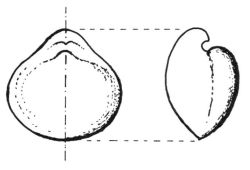

regular or symmetrical, on a centre line drawn from the beak. Then compare the side view with *Protocardia*, opposite.

The Cambrian fossil *Lingulella* burrowed in the sand over 500 million years ago. It was similar in shape and life habit to the living *lingula* or goosebill.

Some brachiopods are called lamp shells, because they are shaped like oil lamps of ancient times and the pedicle hole resembles a wick hole.

Lingulella davisi 190 mm
Cambrian brachiopods

Roman pottery lamp 110 mm
300 AD

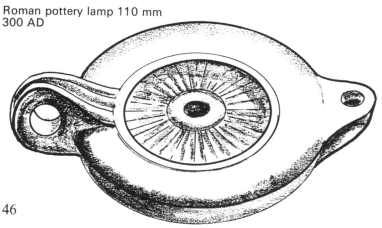

Terebratula species
Jurassic brachiopod

46

MOLLUSCS

Mollusc is a Latin word meaning a soft bodied or boneless animal. There are three important classes in the phylum MOLLUSCA:

1. LAMELLIBRANCHS–oysters, mussels, clams, scallops
2. GASTROPODS–snails, whelks, limpets
3. CEPHALOPODS–squids, nautiluses, the extinct ammonites

Oxytoma cygnipes, Spisula arcuata and *Protocardia truncata* are fossil examples of lamellibranchs (the word refers to the gill structure). They have two shells or valves that form a pair, and so they are often called bivalves. The living creature has a hatchet shaped fleshy foot, and for this reason the lamellibranchs are also known as PELECYPODA, or hatchet feet.

Gastropods are univalves; they have one shell that is carried above the body. The animal can withdraw into the shell, and the entrance is closed by a hard plate, the operculum. Though not common in Palaeozoic rocks, fossil gastropods are abundant in Mesozoic and Caenozoic formations. *Volutaspina* is a gastropod with the coil, or volute, ornamented with spines, and is but one example of the many different fossils of this class. There are others to be seen in this book.

Cephalopods are dealt with in detail on pages 50 to 51.

Oxytoma cygnipes 75 mm
Jurassic lamellibranch

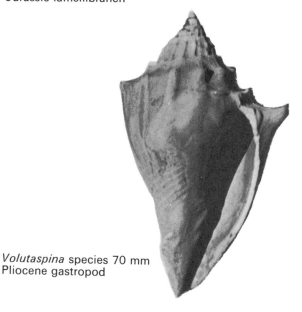

Volutaspina species 70 mm
Pliocene gastropod

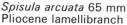

Spisula arcuata 65 mm
Pliocene lamellibranch

Protocardia truncata 30 mm
Jurassic lamellibranch
Lamellibranch valves form a pair

Ophioderma tenuibrachiata 85 mm
Jurassic starfish

ECHINODERMS

The starfish and the sea lily are both sea animals related in the phylum ECHINODERMA (spiny skins). Other classes of echinoderms were shown on page 36. The shape of some of these creatures is based on a five sided figure, the pentagon; there are five arms on the starfish *Ophioderma*. The ossicles or small bones of the sea lily *Pentacrinus* resemble separate pentagonal beads of a broken necklace. Echinoids or sea urchins are common fossils in limestone rocks of the Cretaceous period.

Pentacrinus species 150 mm
Jurassic sea lily

OPERATION
'Neptunea Contraria'

Here two young palaeontologists are seen at work removing a fossil shell from its bed, where it had lain undisturbed for a million years. The bed is of Pleistocene age, and is called the Red Crag, a soft friable (crumbly) rock made of broken shells and sand. It is known to be rich in fossil gastropods and bivalves.

The boy and girl prepared for their expedition by studying a book on Caenozoic fossils and learning the names of those they might hope to find. They therefore knew what to look for. They were specially interested in the gastropod *Neptunea contraria*, an unusual species of whelk with a left hand or sinistral coil, this being contrary to most gastropod spirals.

The method of extracting the fossils was simple. The loose crag was scraped away with a small trowel and a penknife, so that the fossil was fully exposed and could then be lifted out by hand. Some of the shells were fragile and needed careful handling.

Individuals of seven different species were collected by the fossil hunters.

GASTROPODS:
1. *Neptunea contraria*   4 adults, 3 juveniles
2. *Hinia reticosa*   2   ,,
3. *Nucella tetragona*   4   ,,
4. *Emarginula reticulata* 2   ,,
5. *Polinices hemiclausis* 1   ,,

LAMELLIBRANCHS:
1. *Spisula arcuata*   2   ,,
2. *Glycymeris glycymeris*   6   ,,

Two young palaeontologists carefully uncover *Neptunea contraria* 28 mm

Ancient Greek coin with Alexander
the Great as the god Ammon

Goniatites species 60 mm
Carboniferous cephalopod

Cross section of a cephalopod
showing septa 125 mm

CEPHALOPODS

The Cephalopods are the highest class of the phylum MOLLUSCA (see page 47) because they have a distinct head and body, and more advanced nervous systems and digestive organs than other molluscs. Cephalopod means 'head-foot', the tentacles or 'feet' being set around the head in the shell opening.

They are categorized as follows:

Phylum: MOLLUSCA

Class: CEPHALOPODA

Orders: Nautiloids (nautiluses) Ammonoids (goniatites and ammonites), Dibranchia (belemnites)

The ammonoids and belemnites are now extinct. Nautiloids appeared in the Palaeozoic era as the first cephalopods. The fossil shells are found as straight tapered cones, some of giant size up to three metres long.

The spiral curled shell is distinctive of the AMMONITES, which take their name from the ancient god Ammon. On Greek and Roman coins his head appears with ram's horns curling back from the forehead.

GONIATITES were the first ammonoids, and the fossils are easily recognised by the bold chevron, or zig-zag shaped markings – the *suture* ('stitched effect') lines that show the various chambers of the shell. These chambers are revealed here in a cut open section. The last and largest was occupied by the soft body of the animal. The smaller chambers were empty and could be filled with air or water by a tube or *siphuncle*, to act as buoyancy tanks for the submarine ammonoid.

The Jurassic oceans must have swarmed with ammonites, for they are found in great numbers as fossils. They are classified in genus and species according to the number of turns in the spiral, the ribs on the shell, the absence or presence of a keel (outer rim of the coil) and the cross section of the whorl or spiral turn.

Ammonite
Position in life and modelled head

Belemnite
Position in life

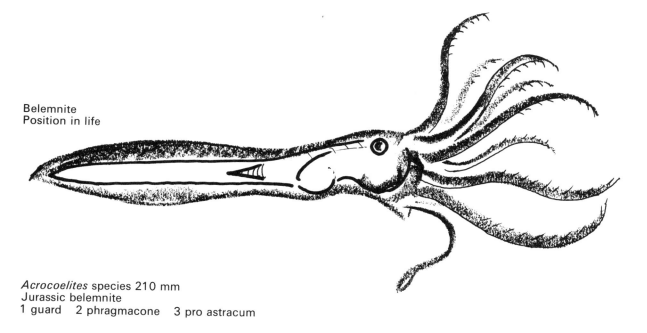

Acrocoelites species 210 mm
Jurassic belemnite
1 guard   2 phragmacone   3 pro astracum

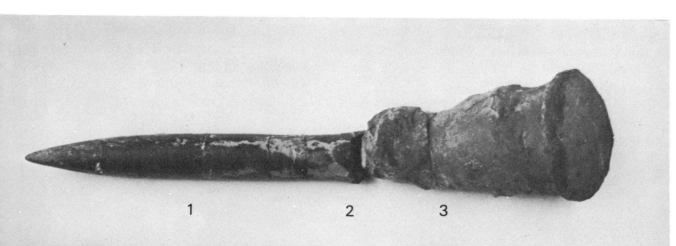

1   2   3

Arisphinctes maximus 480 mm
Jurassic ammonite

Belemnites were like the living cuttle-fish. They did not have a hollow shell, but the fleshy mantle enclosed a hard rod, like a cigar in size and shape, with a small hollow shell, the *phragmacone*, attached. The rod, or *guard*, is preserved as a fossil.

On this giant ammonite part of the shell has been removed, to reveal fine markings like branches of a tree. These are the suture lines, and they show the position of the walls that divide the shell into many separate chambers. The intricate pattern made by the sutures on this ammonite may be compared with those on the Triassic ceratite on page 64, where these are seen to be simple wavy lines.

52

DINOSAURS

Early in the nineteenth century some fossil fragments of bone and teeth were found in England which resembled those of a crocodile. They were recognised as the remains of a DINOSAUR, which means 'the terrible lizard', from the Greek *deinos*–terrible and *sauros*–lizard. Except for the name, however, little more was known of these creatures until many years later, when more discoveries were made in America. Huge fossilized bones were found exposed in the rocks of the badlands in Colorado and Wyoming. The terrifying size of the dinosaurs was then revealed. Two American palaeontologists, Othniel Charles Marsh and Edward Drinker Cope, worked separately and in much rivalry over the years, to find and restore the bones to complete skeletons.

In museums all over the world these dinosaur skeletons now capture the popular imagination as monsters of the prehistoric past. In comic books and films the dinosaur eternally chases the Stone Age man. This is a completely false idea, however, for the great race of dinosaurs died out at the end of the Mesozoic era, long before Man first appeared.

Brontosaurus lived in the swamps and fed on reeds and grass. It was a harmless herbivore, as were many of these giant reptiles. *Triceratops* was a grass eater of the plains, equipped with defensive armour of bony plates and horns. The fierce *Tyrannosaurus* was a flesh eating animal, a carnivore. With sharp teeth set in large powerful jaws, it is well named 'the tyrant dinosaur'.

These all lived in the Jurassic and Cretaceous periods.

Herbivores: (length from head to tail)

Brontosaurus (thunder lizard)	20 metres
Diplodocus (double beam bone)	25 metres
Cetiosaurus (whale lizard)	18 metres
Stegosaurus (covered lizard)	6 metres
Triceratops (three horned)	7 metres

Carnivores:

Tyrannosaurus (tyrant lizard)	15 metres
Ceratosaurus (horned lizard)	5 metres

Brontosaurus species
Painting by a junior artist

The dinosaurs existed and survived over a period of a hundred million years and then they died out quite quickly at the end of the Cretaceous. Why did they become extinct? It is said that because they were so big they became clumsy and helpless, unable to look after themselves. But this does not explain why they lasted so long as the leading reptiles of the Mesozoic Era. It is more probable that they dwindled in numbers and died from starvation and cold. At the end of Cretaceous times there were changes in land surfaces through orogenic movement (mountain building) and the climate became colder. The warm swamps, the home of the dinosaurs, dried out and became cold uplands. Life no longer favoured the dinosaurs. It was now the turn of the mammals.

Tyrannosaurus Rex
Cretaceous carnivore reptile

Triceratops species
Cretaceous dinosaur — length 6 metres

Cryptocleidus species postage stamp
Jurassic plesiosaur

MARY ANNING

In southern England the Jurassic rocks out-crop on the coast at Lyme Regis. The locality is famous for the abundant fossils contained in the Blue Lias rocks, the scene of some exciting finds made by a young palaeontologist called Mary Anning. She was born in Lyme Regis in 1799, and as a child helped to support her family by collecting ammonites and shells, to sell as curios to the summer visitors. Long hours of work spent in this way gave her practical experience of fossils, of more value than book learning and museum study. At the age of twelve she found and dug out a complete skeleton of what looked like a giant fish. This proved to be not a fish but a reptile, *Ichthyosaurus*, the fish-lizard. It was the first whole specimen of this fossil, former-ly known only by odd fragments of bone. Mary then became well known as a supplier of fossils to museums and learned men, and as an expert on Jurassic specimens. She made many more important finds, such as the sea reptile *Plesiosaurus* and the flying reptile *Dimorphodon*. Her success in this work lay not in strokes of luck but in keen observation and knowledge of ancient life, as fossils are readily seen in the rocks by trained eyes.

Ichthyosaurs first appeared in the Triassic period, and then became extinct at the end of the Cretaceous. They were like our modern dolphins in appearance, although they were reptiles and the dolphin is a mammal. They had streamlined bodies and four strong pad-dles, for smooth and rapid movement in the water. These paddles were not the delicate fins of fish but limbs of bone and muscle. Fossils of these large reptiles are sometimes found with small ichthyosaurs inside them, which suggests that the young were born as babies resembling the adult, and not hatched out like the chicken from the egg. The female ichthyo-saur could give live birth to her young at sea, and did not lay eggs in the sand as do many marine reptiles. Like the mammals, they were *Viviparous*–giving live birth to young.

Mesozoic sea reptiles:

Ichthyosaurus (fish lizard)	9 metres
Plesiosaurus (near lizard)	12 metres
Elasmosaurus (plate lizard)	
cryptocleidus	14 metres

Mary Anning discovers the first
plesiosaur 1824

'Northern Echo' photograph

OPERATION *'Ichthyosaur'*

As you will see, this boulder rests on the sea shore, as there are living limpets attached to it, which also indicates that it would be under water at high tide. Looking at it closely, you will observe a certain regularity in the shape and some differing textures on the surface. These could be traced in outline to the shape of a head with a large eye socket. It is the fossil skull of the fish-lizard, *Ichthyosaurus*.

The forepart or snout is missing, and the drawing opposite is a reconstruction of the whole skull. A group of geologists found this large fossil on the rocky foreshore, below highwater mark, where it may have lain for some time after a cliff fall. An important fossil such as this needs expert handling, and you should report finds of this sort to your nearest museum.

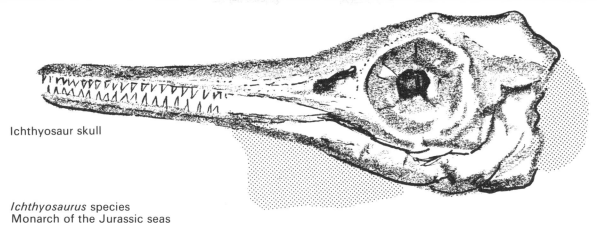

Ichthyosaur skull

Ichthyosaurus species
Monarch of the Jurassic seas

Plesiosaur
Jurassic sea reptile

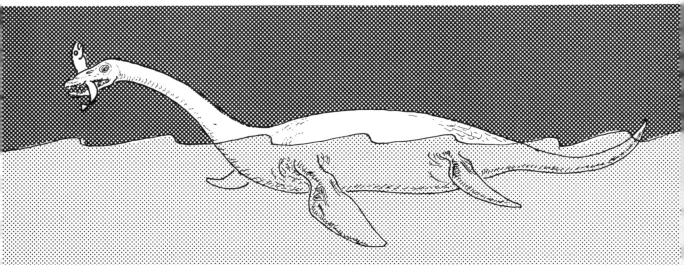

EARLY BIRD

This fossil skeleton of a little bird is more important to paleontologists than a dozen dinosaurs. *Archaeopteryx,* for instance, is a particularly rare and distinguished fossil. Only three specimens have been found, and they show that there is a link between reptiles and birds in the succession of life. The earliest birds had a reptilian bone structure, the skull set with teeth, and the backbone extending to a long bony tail; but they were also completely fledged, the feathers having evolved from reptile scales.

These fossils came from Jurassic formations in Bavaria, where a fine limestone is quarried that was once the bed of a lagoon. In the nineteenth century this fine grained stone was used in the newly discovered process of lithography, printing from stone. While the stone was quarried for this purpose, it also yielded excellent fossils of plants, shells and fish. In 1861 an unusual specimen was unearthed, the clear impression of a feather 68 mm long. This was a startling find. Flying reptiles were known, but none had feathers. Even more exciting was the discovery, in the same year, of a complete skeleton with wing and tail feathers clearly imprinted in the stone – the earliest known bird. This unique fossil was acquired by the British Museum. A second specimen was found in 1877 and a third in 1958. Perhaps more will come to light.

Archaeopteryx lithographica, the ancient bird of the lithographic stone, was not a powerful flyer, for the reptilian bones and the clumsy tail would not be suited for sustained flight. It is classified as a bird:

Class: AVES (birds)
Sub-class: ARCHAEORNITHES (ancient birds)
Order: Archaeopterygiformes
   (ancient feathered birds)
Genus: *Archaeopteryx*
Species: *lithographica*

Pteranodon species
Cretaceous flying reptile – wing
span up to 8 metres

Archaeopteryx lithographica
250 mm
Jurassic bird

Oreopithecus species
Pliocene ape
British Museum photograph

THE PRIMATES AND MAN

There is something dramatic and appealing in this fossil skeleton. The sprawling pose suggests a recent human tragedy, rather than a remote and impersonal link with the past. The bones appear human, but in fact they are of a fossil ape, *Oreopithecus*—the mountain ape. This beautiful specimen was recovered from a coal mine in Italy in 1958. It is of Pliocene age, which makes it about ten million years old, much earlier than any known human fossil remains. The likeness to human bones, especially those of the skull, first suggested a link with Man, but opinion is now against this. The ape is not our direct ancestor; he is a distant cousin.

In Mesozoic times some primitive mammals of the forest clawed their way up the tree trunks, to live in safety among the branches. They fed on insects and tender leaves and shoots, and adapted their limbs to climbing and swinging among the branches. So they developed powerful arms and grasping hands from the forelimbs of the quadrupeds. The senses of hearing, smell and sight became sharpened, and the pace of life quickened with these intelligent little mammals; all this in contrast to the slow moving reptiles, then the dominant race of the Mesozoic era. From these first mammals came the PRIMATES, an order which includes lemurs, monkeys, apes and Man himself.

Fossils of land animals are less common than those of the sea, and it is not surprising that the remains of early man are rare. They are usually fragments of the skull, with teeth, and parts of the long bones. From this scanty evidence the prehistory of Man has been recorded.

Early manlike apes, *pithecanthropoids*, left the forest to hunt the animals of the open grasslands. They lived in caves, they stood erect and their legs and feet were strong for running.

Skulls and bones discovered as fossils in China and Java have been classed as true early men. They are *Sinanthropus pekinensis* (Peking man) and *Pithecanthropus erectus* (Java man). From these finds it seems that Man originated in Asia.

In Europe, early man is represented by *Homo neanderthalensis*, called Neanderthal man from the place in Germany where his bones were unearthed, in 1856. More remains of Neanderthal man have been found, and these first Europeans are now seen as men of stocky build, with low brows and strong jaws. They lived in family groups and used simple flint tools.

The race of Neanderthalers died away in the fourth and last great Ice Age of the Pleistocene. They were replaced by a new human type, whom we recognise as *Homo sapiens* – modern man. First evidence of this new race was found in a cave at Cro-Magnon, in France, where five skeletons were preserved, dating back thirty thousand years. Cro-Magnon man is our European ancestor.

Homo neanderthalis
Primitive European man

Homo sapiens
Cro-Magnon man
Early Stone Age

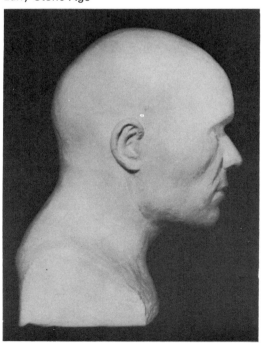

Ceratites nodosus 130 mm
Triassic cephalopod—note the
sutures

FOSSIL HUNTER OR PALAEONTOLOGIST?

Twenty different species of ammonites are shown and named in this book, but these are only a few of many hundreds known and classified. This profusion of species is to be found in all the shellfish classes and in corals, plants and all common fossils. With all this material at hand there is a lot of work which could be done by individuals and by groups, working in the spirit of scientific inquiry.

When you search for fossils, you are a fossil hunter. But when you search for knowledge about the fossils, you are then a palaeontologist. Part of your research work can be simply measuring and counting. For instance, how many different fossils are found in a square metre of rock surface? Which is the commonest fossil, and what is the average size of the species, measured in length by millimetres? There are endless more questions like these to be answered.